MW01359053

Written by Nathalie Tordjman
Illustrated by Laura Bour

Specialist Adviser:
Roderick G. Christie,
Director of Education
Teatown Lake Reservation (New York)

ISBN 0-944589-38-3
First U.S. Publication 1991 by
Young Discovery Library
217 Main St. • Ossining, NY 10562

©Editions Gallimard, 1989
Translated by Vicki Bogard
English text © Young Discovery Library

The Living Pond

YOUNG DISCOVERY LIBRARY

If you step up to the edge of a pond, you will see a disappearing act! Some animals dive under the water. Others fly away squawking or buzzing.

Behind the tall grass, there's a lot going on! Animals live around, on, and under the water. And many creatures, just passing by, benefit from a pond.

If the banks or sides of the pond are steep, not many water plants will grow there.

The land is full of high and low places. When rain fills up a low place, a pond is born. The water stays there—it is **stagnant**. The same animals live in pools, ponds, and lakes. But a pool is smaller than a pond. It dries up in the summer and the animals move away or die. A lake is bigger, but it has less plants growing in it than a pond.

If the banks of a pond slope gently, more kinds of water plants can grow there.

In the Middle Ages, people dug ponds to raise fish.

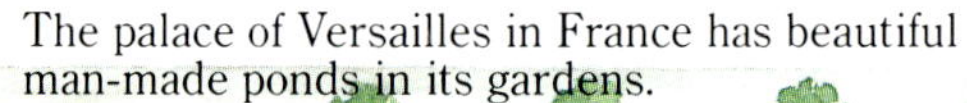

The palace of Versailles in France has beautiful man-made ponds in its gardens.

Old abandoned sandpits turn into ponds.

Ponds used for swimming and boating don't have many animals in them.

A large rectangle with steep sides makes a good fish pond.

Some ponds are natural.
They started as hollows in the ground. Rainwater filled them up and stayed there. The bottom of the pond is full of bumps and holes—good for plants to grow.

First, plants grow around the sides with their "feet" in the water.

Many ponds are man-made.
People leave holes in the ground when they dig out sand, or gravel for a driveway. After a few years, the holes look like natural ponds.

When plants die, they pile up around the pond. Other plants then grow in their place.

Others are created when farmers connect small pools to make one big pond. The farmers keep the ponds clean and use them to store water or for raising fish.

Small shrubs begin to grow.

Many water birds make their homes on ponds.
Ducks and grebes have special feet so they can paddle fast on water. Grebes dive to catch small fish. Pond ducks stay on the surface, but they can fly straight up in the air if there is danger from above or below the water.

The dragonfly darts over
the water, gobbling other insects.
It lays its eggs on the
stems of water plants.

On the water's surface,
you can see some champion skaters.
Water striders zip along on their
hairy legs and don't even get wet!
Whirligigs spin their
oily bodies around
in circles.

How do insects move under the water?

Most do the breaststroke, except for the backswimmer, of course! Its long back legs look like feathery oars. Diving beetles use their hind legs to kick. They also have wings and can fly away to another pond, if they want to.

Frogs lay their eggs underwater and the eggs develop into tadpoles. Young tadpoles have no legs. They use their tails to swim, just like fish. Crayfish do have legs—they scuttle along on the bottom. They can also jump backwards by flicking their tails hard.

The kingfisher spots a fish from high up on its perch. Then it dives headfirst into the water to grab it.

minnow
carp

A pond has an underwater forest. Plants cover most of the pond's bottom. Sunlight reaches down into the calm, shallow water. The plants give **aquatic** animals oxygen, food and shelter. **Herbivorous** (plant-eating) animals keep the plants from taking over. So, every plant and creature has a part in the life of a pond.

Freshwater snails nibble on **algae** that grow on plants and rocks.
Tadpoles are not very fussy. When they are young, they eat plants. When they get older, they eat dead fish, or even other tadpoles!
Fish such as carp eat plants, but they also like insects, mudworms and small crayfish. (Crayfish look like little lobsters.)

Freshwater clams live in the mud. They feed on very small, almost invisible plants that they filter from the water.
Tiny snails, crabs and mudworms keep the bottom clean by chewing up dead plants.

In the pond, the prey always outnumbers the hunters. This large pike has few enemies… only a heron or a fisherman could catch it.

The newt eats small animals.

What is that pike doing hidden in the plants? It is waiting for a smaller fish, a frog, or a coot chick to go by. Then in a flash, it grabs its prey.

Diving beetles find a lot of tasty snacks during their swims!

This mud turtle is a rare and threatened species.

It competes with the water shrew for food. They both like the insects and tadpoles that hide among the plants and around rocks.

The water spider
sets up house under the water to surprise its prey. It fills its web with air bubbles. From this shelter, the spider grabs any insects or tadpoles that pass by its door.

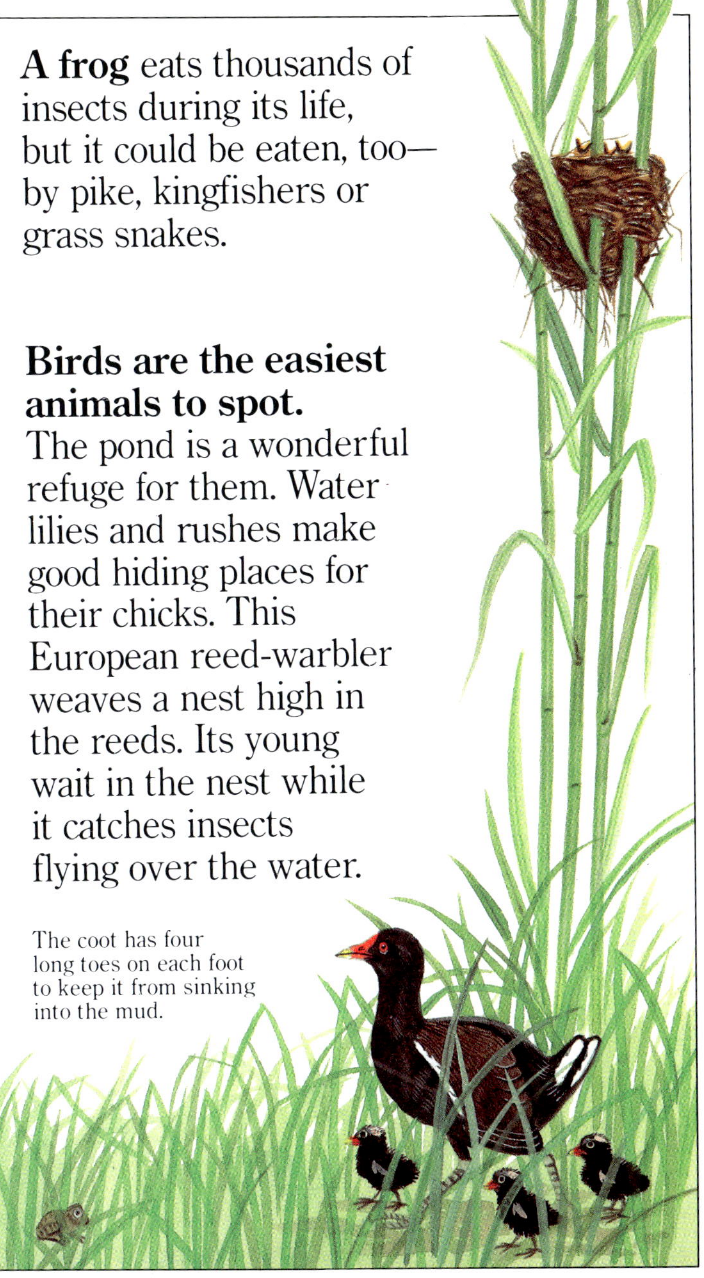

A frog eats thousands of insects during its life, but it could be eaten, too—by pike, kingfishers or grass snakes.

Birds are the easiest animals to spot. The pond is a wonderful refuge for them. Water lilies and rushes make good hiding places for their chicks. This European reed-warbler weaves a nest high in the reeds. Its young wait in the nest while it catches insects flying over the water.

The coot has four long toes on each foot to keep it from sinking into the mud.

Like all living things, aquatic animals need oxygen. Land animals take oxygen from the air with their lungs. Fish take oxygen from the water with their gills.

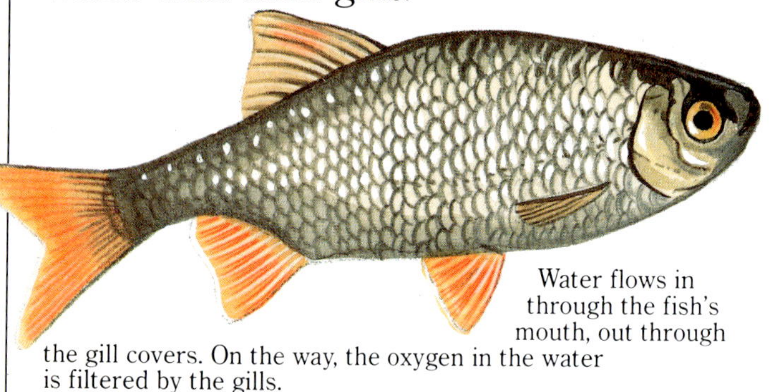

Water flows in through the fish's mouth, out through the gill covers. On the way, the oxygen in the water is filtered by the gills.

Freshwater clams also breathe by filtering water through their gills.

Crayfish have gills that look like feathers. They are under the shell in the same part of the body where your lungs are—the thorax, or chest.

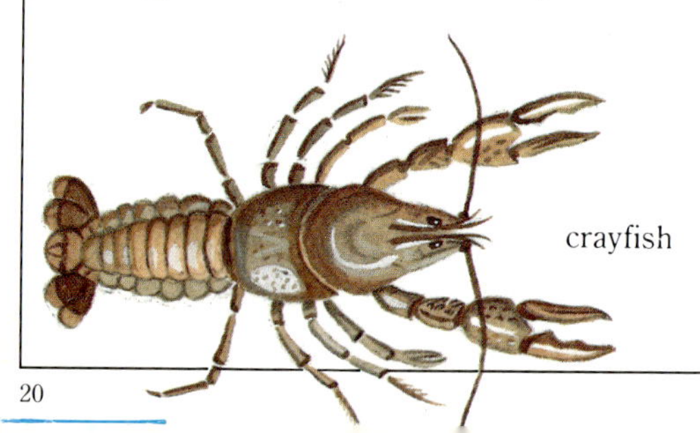

crayfish

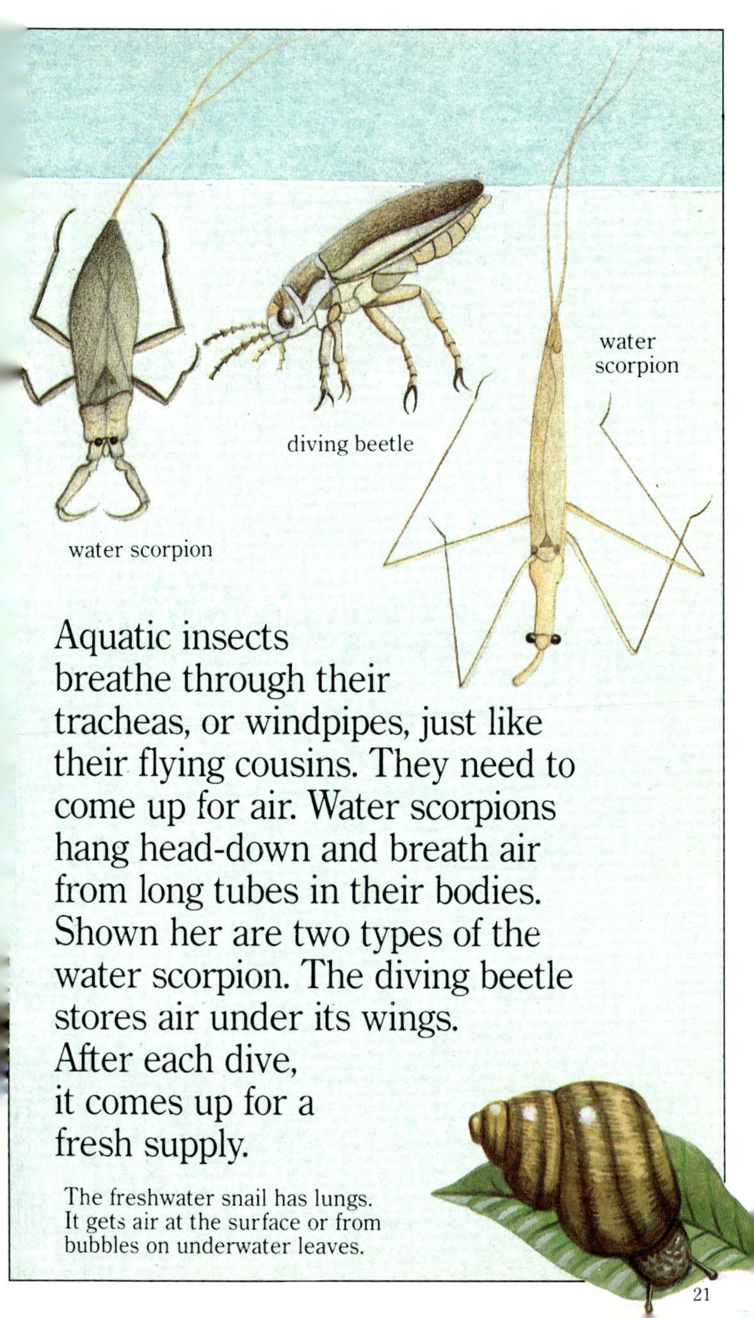

Aquatic insects breathe through their tracheas, or windpipes, just like their flying cousins. They need to come up for air. Water scorpions hang head-down and breath air from long tubes in their bodies. Shown her are two types of the water scorpion. The diving beetle stores air under its wings. After each dive, it comes up for a fresh supply.

The freshwater snail has lungs. It gets air at the surface or from bubbles on underwater leaves.

In the spring, the pond is a real nursery!

The grebe makes a floating nest for its family in the middle of the pond.

The kingfisher digs a long tunnel in a steep bank. Then it builds a nest at the end and lays its eggs.

After the grebe's eggs hatch, the chicks snuggle down on their mother's back. Turtles lay their eggs on the banks of the pond. When the babies hatch, they head straight for the water. Dragon-flies do just the opposite! Their eggs hatch in the pond. The **larvae**, or nymphs, grow underwater. When they become adults, they climb onto a reed or a rock. They then shed their skin and fly away on their new wings.

The muskrat feeds her babies with milk. Her teats are high up so she can nurse and swim at the same time.

Frogs, toads and salamanders live on land much of the time, but in the springtime, they lay their eggs in calm, still water.

Salamanders mate in the water. Then the female hangs her eggs on the leaves of aquatic plants.

Most fish do not mate. After a courting dance, the female lays her eggs. Then the male covers them with milt, or semen.

Most fish spend their whole lives underwater. Some make nests for their eggs and guard them until they hatch. Others just lay their eggs and swim away.

A baby salamander breathes through its gills like a fish. When it becomes an adult it has developed lungs.

Young fish are not shy. They live in little groups and come right up to the edge of the pond.

By the time mosquito larvae hatch, their parents are already far away. Tadpoles also have to make it on their own. Only a few of them reach adulthood because they are eaten by other animals.

Mayfly larvae spend a year or two in the pond. They have gills on their abdomens.

Damselfly nymphs and diving beetle larvae eat mudworms, tadpoles, even small fish. Caddisfly larvae are called caddisworms. They make cases to live in from their own silk and wood or sand.

tadpoles

damselfly nymph

They are plant eaters. By the time they become adults with wings, they only have a few hours left to live!

During northern winters, the pond freezes over. The fish move to the bottom, away from the frost. Insect larvae and salamanders **hibernate** in the mud. Hibernation is a long, deep sleep.

In the summer, the surface of the pond heats up fast. Animals dive to the bottom to stay cool. Algae grows fast and makes the water look very green.

Many animals need the pond to survive. Today, ponds are often drained. Others are visited by noisy people who scare animals away. Sometimes the water is polluted. Then the pond is not a good place for animals to live in.

Long ago, people thought ponds brought bad luck and disease. Some even thought that witches or mermaids lived there!

Farmers drain ponds to make more room for their crops. But after a few years, the harvest is not as good as before. Soon, only grass and weeds will grow there.
Near cities, ponds are sometimes turned into places for boating and swimming. The plants that grow around the pond are pulled up to make room for a beach. Then the animals move away or die.

Sometimes a pond will disappear all by itself. Water plants die. Then there is not enough oxygen for the animals to breathe. They cannot live there.

Industrial waste, dumped into ponds, will poison the plants and animals.

The jacana, or lily-trotter, with its long toes and claws, uses water lilies like stepping stones.

Tropical forests are very damp. It rains often. Rivers overflow their banks and flood the forest. Then there are huge puddles of stagnant water. Crocodiles hide there, stalking their prey. Only their nostrils stick out of the water. They catch fish or grab birds and mammals that swim on the water's surface.

The capybara is the world's largest living rodent—two feet high and three to four feet long. It lives in South America. Parts of its feet are webbed so it won't sink in mud.

Teeny tiny animals!

Even if you do not live near a pond, you can take water from a puddle. Even a bucket full of rainwater, or a vase with flowers that are dead, will do for this experiment. Put a drop of water under a microscope or a big magnifying glass. You will see algae and tiny animals with see-through bodies.

How can you see what is going on under the water? Gently dip a net in the water and scrape the bottom. Pull the net out, taking care not to stir the mud.

You can make your own net out of a forked branch and a mesh or burlap bag.

If you find some diving beetle larvae or dragonfly nymphs, don't put them in the same jar with ranatras, water striders, backswimmers or water scorpions. They will eat them! And be careful—some insects bite. Please do not keep them too long in the jar. When you are done observing them, put them back where you found them.

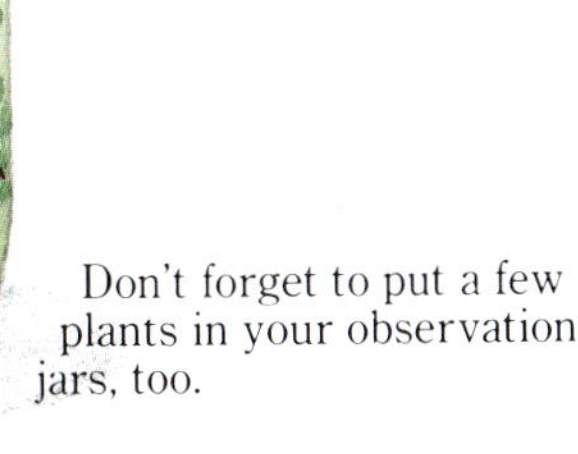

Don't forget to put a few plants in your observation jars, too.

TWENTY FROGGIES

by George Coope

Twenty froggies went to school
Down beside a rushy pool
Twenty little coats of green,
Twenty vests all white and clean.

"We must be in time," said they,
"First we study, then we play;
That is how we keep the rule,
When we froggies go to school."

Master Bullfrog, brave and stern,
Called his classes in their turn,
Taught them how to nobly strive,
Also how to leap and dive;

Taught them how to dodge a blow,
From the sticks that bad boys throw.
Twenty froggies grew up fast,
Bullfrogs they became at last;

Polished in a high degree,
As each froggie ought to be,
Now they sit on other logs,
Teaching other little frogs.

Index